Cambridge Industrial and Commercial Series
General Editor: G. F. BOSWORTH, F.R.G.S.

SHIPS, SHIPPING AND FISHING

SHIPS, SHIPPING AND FISHING

WITH SOME ACCOUNT OF OUR SEAPORTS AND THEIR INDUSTRIES

by

GEORGE F. BOSWORTH, F.R.G.S

Cambridge :
at the University Press
1915

CAMBRIDGE UNIVERSITY PRESS
Cambridge, New York, Melbourne, Madrid, Cape Town,
Singapore, São Paulo, Delhi, Tokyo, Mexico City

Cambridge University Press
The Edinburgh Building, Cambridge CB2 8RU, UK

Published in the United States of America by Cambridge University Press, New York

www.cambridge.org
Information on this title: www.cambridge.org/9781107693869

First published 1915
First paperback edition 2011

A catalogue record for this publication is available from the British Library

ISBN 978-1-107-69386-9 Paperback

GENERAL PREFACE

THE books in this Series deal with the industrial and commercial condition of our country. Of the importance of the subject there can be no doubt, for it is the story of the material side of the life of a great nation. British agriculture is the most enterprising in the world; British manufactures, both textile and hardware, are famed in all parts of the globe; British ships are on every sea and carry for other nations as well as for ourselves; and Britain, through the Banks and Exchanges of London, is the centre of the money market of the world.

It has been well said that material needs cannot be neglected or forgotten with impunity in this world. Just as a man must have bread to eat if he wishes to enjoy life, so a nation needs material prosperity if it is to be of real influence in the world. Industrial and commercial prosperity does not, in itself, constitute greatness, but it is a condition without which national greatness is impossible. Hence, the story of the industrial and commercial condition of Britain is worth telling to our school children, not only that they may rejoice in our country's progress, but, also, that they may realise the responsibilities borne by the citizens of the first of all nations.

G. F. B.

EDITOR'S NOTE

THIS first book of this Series deals with Ships, Shipping
and Fishing, and should be popular in all our schools,
more especially those round our coasts. It traces the
development of man's ingenuity from the early dug-out
boat to the launching of the *Aquitania*, and it gives a
succinct account of the work of the Royal Navy and the
Dockyards. The Fishing Industry and Fishing Ports are
specially considered, and such well-known institutions as
Trinity House and "Lloyd's" are described. Lighthouses
and Lightships, their construction, distribution, and value
to the shipping industry are explained; and our great
Seaports and their industries, together with the past and
present condition of the Cinque Ports are treated in the
last chapters.

Teachers and students who wish to study in greater
detail any part of the Industrial and Commercial History
are advised to use Dr Cunningham's *Growth of English
Industry and Commerce*, published by the Cambridge
University Press. There they will find full and accurate
references to a large number of authorities on all branches
of this subject.

G. F. B.

August 1915

CONTENTS

MAPS AND ILLUSTRATIONS

The illustration on p. 4 is reproduced by courtesy of Mr Cecil Torr; that on p. 14 by courtesy of the Cunard Steamship Co.; that on p. 13 is from a block kindly lent by Messrs J. Maxwell & Son, Dumfries.

1. EARLY BOATS AND SHIPS

THE MEDITERRANEAN

The trunk of a tree floating down a river may have given early man the first notions of water carriage. It would not be long before he discovered that the tree could support more than its own weight without sinking. Then would follow the knowledge that several trunks fastened together forming a raft would carry men and goods on the water. Even to the present day this mode of water carriage has continued and may be seen in various parts of the world—on the Rhine, on the rivers of Canada, and on the Euphrates and Tigris.

As men found the floating and carrying capacity of a log or trunk, the idea would soon develop that the weight of the log or trunk might be lessened by hollowing it out and thus increase its carrying capacity at the same time. Here we arrive at the first crude idea of the modern boat. The early hollowed trunk of a tree or the "dug-out" was used by the prehistoric people of Britain, and specimens have been unearthed in recent years in the beds of rivers and other places.

In course of time improvements were made in the portion under water so as to make the boat handy and offer less resistance to the water: then followed, as the result of experience, better accommodation in the interior and more facilities for navigating the craft. Such boats or canoes are now made by many uncivilised people, such as the islanders of the Pacific and some of the tribes of Central Africa.

The built-up boat followed and was often a mere framework of bone or wooden ribs covered with hides or tree bark. The coracle of the early Britons was of this type and this class of boat is still used on the west coast of Ireland. The change from a raft to a flat-bottomed boat or coracle was a great improvement; but it probably took a greater interval of time to evolve decked ships and the date of their appearance is very uncertain.

It is sometimes stated that Noah's ark is the oldest ship of which there is any record; but pictures of vessels

Oak canoe found in the Tyne

have been discovered much older than the ark and there are vessels now in existence in Egypt which were built before the period when Noah's ark was constructed. In 1894 a discovery was made of several Nile boats of the time represented by 2850 B.C., admirably preserved in brick vaults, a little above Cairo, on the left bank of the river. The boats are about 33 feet long, 7 to 8 feet wide, and 2 to 3 feet deep, and as they were found near some royal tombs it is considered probable that they were funeral boats used to carry royal mummies across the river. They are constructed of planks of acacia and sycamore, and were steered by two large paddles. Although they are nearly 5000 years old, the method of construction was so satisfactory that they held firmly together after their supports had been removed.

After the Egyptians, the Phoenicians were among the first of the people who dwelt on the Mediterranean seaboard to make progress in sea-going ships. This remarkable people lived along the seaboard to the north of Palestine and had Tyre and Sidon for their chief cities. Their ships were of considerable size, well-manned and equipped, and traded with all parts of the then known world. They made voyages to Cornwall for tin, to the south coast of Ireland, and along the western shores of Africa. Not only were the Phoenicians a great trading people themselves, but they built ships for other people and even manned the fleets of the Egyptians and Persians. There are a few representations of the Phoenicians' ships, and one of them shows us that the warship was a galley carrying a single square sail and with two banks of oars on each side. The galley had a ram and the steering was by means of two large oars at the stern. The soldiers were carried on a deck or platform raised above the heads of the rowers.

Let us now pass to the time of ancient Greece and Rome.

First we must notice the favourable natural conditions that existed in Greece for the improvement of shipbuilding. The mainland on the east and west was bordered by inland seas studded with islands having good harbours; and one of the chief cities, Corinth, had a unique position for trade and colonisation. The Greeks had both the commercial and colonising instincts, and the benefit of the maritime experience of the Egyptians

and Phoenicians. It was not till the fourth century B.C.
that they appear to have made any improvements in
ships. The earliest naval expedition recorded in Greek
history is that which carried the soldiers to the siege of
Troy, about 1200 B.C., and the historians tell us that the
vessels used were open boats.

In the British Museum there is an Athenian painted
vase having an illustration of a single-banked Greek
galley of about 500 B.C. The vessel is armed with a
ram and has seventeen oars on each side. The mast and
rigging are indicated by lines. From other Greek vases
we gather the knowledge that soldiers fought on the deck
proper and on a raised forecastle. At a later period the

Ancient ship on vase

Athenians built ships with four and five banks of oars,
and Alexander the Great is said to have used ships of the
latter type.

The later merchant ships of the Mediterranean nations
did not differ greatly from the warships of the time,
although there were more distinctions among those of
the ancient Greeks and Romans. About 256 B.C. the
Romans became a naval power, for the command of the
sea became a necessity of their existence. Their first
fleet seems to have been of the same type as the Greek
galleys; and their many-banked ships appear to have
been large and unmanageable. In time these were
superseded by the swift, low, and handy galleys, and it

was with such vessels that Augustus gained the battle of Actium in 31 B.C. Henceforward this type was used for Roman warships and the word "trireme" came to signify a warship, without reference to the number of banks of oars.

After the Romans had completed the conquest of the nations bordering on the Mediterranean, naval war ceased for awhile and the Roman navy declined in importance. The galley, although somewhat modified in course of time, continued to be the principal type of warship in the Mediterranean till about the sixteenth century.

2. EARLY BOATS AND SHIPS

THE NORTH SEA

The preceding chapter was concerned with the development of ships on the coasts of the Mediterranean Sea. Let us now turn our attention to the countries bordering on the North Sea—the sea which was the centre of maritime activity in the seventh and eighth centuries of the Christian era, and has again become, in the twentieth century, the scene of a tremendous conflict for naval supremacy.

As early as Caesar's time some of the northern nations had become skilled in the arts of shipbuilding and seamanship. Caesar gives a general description of the ships of the people who lived in the district now known as Brittany. These people had the carrying trade between Gaul and Britain. Their ships were built entirely of oak and placed no reliance on oars, but used sails which were of dressed hide. The bottoms of the vessels were flat so as to be easily beached, and the cables were iron chains. From Caesar's description of these Gaulish vessels it is evident that they were very different from the oar-propelled galleys of the Mediterranean.

At the time of Caesar's invasion of Britain we read that the art of ship construction had made little advance among the Britons, for their vessels were made of a framework of light timber or of wicker, over which was stretched a covering of strong hides.

The Saxons, who invaded Britain in the fifth century, were the pirates of the North Sea, and although they used similar boats to those of the Britons, they were larger and of stronger construction. It was at this period of the Saxon invasion of Britain and their subsequent settlement in our land that the Norsemen were making themselves felt all along the European seaboard. Their fleets entered the Mediterranean on the south and their seamen discovered and colonised Iceland and Greenland. We must therefore consider the Norsemen as the most famous navigators before the period known as the Middle Ages.

It was the custom of the Norsemen to bury their great chiefs in one of their ships and to heap earth over the whole mass. Some of these ship-tombs have been discovered in recent years and from the remains we are able to form a tolerably good idea of the smaller class of vessels used by the Northmen or Vikings. The most important of these old vessels was found at Gokstad in Norway in 1880 and is probably a good representative of the larger type of vessels that went on distant expeditions.

The Gokstad vessel is about 78 feet long, over 16 feet in width, and $5\frac{3}{4}$ feet deep. It was propelled by sails as well as oars, and there were arrangements for raising and

The Gokstad ship

lowering the single mast which carried a single square sail. The rudder, something like a short oar, was worked by means of a tiller fitted into the socket at the upper end. The lines of the vessel were very fine and the boat was probably capable of high speed.

Vessels of the same type were built in Denmark at a very early date, and it is a matter of history that our Saxon kings had considerable fleets to protect our shores from the attacks of the piratical Danes. Alfred the Great, who became king in 871, may be regarded as the founder of the English navy, for he built better and larger ships than those of the Danish invaders. The *English Chronicle* says that his ships were twice as long as the others; some had 60 oars and some had more; they were swifter and steadier and also higher, and were most efficient. It was with a fleet of these new ships that Alfred met and defeated the Danes in 897.

There is a sketch in the British Museum of an Anglo-Saxon ship of this period and from it we get some idea of an early English war vessel. The sail was the chief means of propulsion and the steering was by two large oars. The keel was prolonged to form a ram and there was some sort of structure on deck.

When William the Conqueror invaded our land, he transported his army across the Channel in a fleet of small vessels which were built for this purpose, and which were burnt by his order when his troops had landed. We have illustrations of these ships from the Bayeux tapestry and they are evidently of the type of ship used by the Vikings. Although the ships of the eleventh and twelfth centuries were generally of small size, yet we meet with notices of vessels that were really large. For instance *La Blanche Nef*, commonly called the *White*

Ship, was built for Prince William, son of Henry I. This was a fifty-oared ship and carried 300 passengers and crew. Galley ships of this type continued to be used in European navies till the seventeenth century.

There is no doubt that the Crusades gave a great impetus to shipbuilding and this was one of the causes of the rise of Venice till it became the foremost maritime power. After the Crusades the trade and shipping of England largely increased and a historian of the thirteenth century writes thus: "Oh England, whose ancient glory is renowned among all nations, like the pride of the Chaldeans: the ships of Tarsis could not compare with thy ships; they bring from all the quarters of the world aromatic spices and all the most precious things of the universe: the sea is thy wall, and thy ports are as the gates of a strong and well-furnished castle." We must also refer to the mariner's compass as another cause which contributed to the development of navigation and shipbuilding, for it was introduced into Western Europe during the first half of the thirteenth century.

We cannot do more than refer to the naval activity of Edward III when he invaded France with upwards of 1000 ships, which fact shows that our resources for shipbuilding must then have been considerable. In 1415 Henry V invaded France with 1400 ships, and it was this invasion which resulted in the capture of Harfleur and the victory of Agincourt.

After the reign of Henry V English commerce and shipbuilding made little progress, but a new maritime power, Portugal, was arising, and for some time the Portuguese made great discoveries and extended their commerce. In the meantime, owing to the Wars of the

Roses, English commerce and shipbuilding went from bad to worse, and it was not till the reign of Henry VII that a change was brought about. It was in his reign that the Portuguese discovered the Cape of Good Hope and the route to India, that Columbus discovered America, and that Cabot discovered Newfoundland. The discoveries of Columbus, Vasco da Gama, Cabot, and other navigators mark a new epoch, for hitherto sailors had kept to the great inland seas of Europe or had engaged in coasting trade. Henceforth ships were to penetrate every sea, and trade and commerce were to be world-wide. And in all these new developments England's ships and Englishmen were to take the foremost place. The first two Tudor kings not only encouraged commerce and voyages of discovery, but also paid attention to the needs of the Royal Navy, so that when the Armada came in 1588 our English ships were numerous and equal to the great occasion.

In the reign of Charles I England occupied a high place in the art of shipbuilding, owing to the patronage of the king and to the great superiority of English oak to other European timbers. England's pre-eminence as a maritime nation was strongly marked during the Commonwealth, and all through our naval history of the eighteenth century we were achieving the success which culminated at Trafalgar in 1805. From that year to the opening of the twentieth century, Britain was without a serious rival on the high seas.

3. FROM SAILING SHIPS TO STEAMSHIPS

During the great struggle with Napoleon British maritime commerce more than held its own, and at its close in 1815 Britain was practically the only carrying power in Europe. Through the whole of the nineteenth century down to the present time it may be noted that, as the naval power of our country was increased, so was the commerce extended and the mercantile marine developed.

At the opening of the nineteenth century the East India Company was by far the largest shipowner in the country, and the vessels, which afterwards became famous for their exploits, were called East Indiamen. They were designed to serve as passenger ships, freight carriers, and men-of-war, and in the latter capacity they fought many actions and gained many victories. They were expensive ships to build, and their crews were four times as numerous as are required for modern merchant sailing ships of similar size. As an example we may mention that one of them, the *Thames*, built in 1819, was of 1360 tons, carried 26 guns, and had a crew of 130 men.

The history of the development of sailing ships in the nineteenth century is of deep interest, and it may be recorded here that the Americans made some of the greatest improvements in mercantile sailing ships. They built "clippers" of greater size, increased their speed, and reduced the crew necessary to work them. Some of the American sailing packets traded with British ports and attained a speed of as much as eighteen knots[1] an hour.

---

[1] A knot is a nautical mile or 6080 feet.

It was in the year 1850 that Englishmen determined to build sailing ships which should not only rival but surpass those of the Americans, and in 1856 the *Lord of the Isles*, built at Greenock, beat two of the fastest American clippers in a race to this country from China. From that time onward British merchant vessels gradually regained their ascendency in a trade which the Americans had almost made their own.

It was not, however, by wooden sailing ships that Britain was to secure the largest portion of the world's carrying trade. Two revolutions were making their influence felt in the shipbuilding world : the first was the means of propulsion, and the second was the introduction of iron instead of wood for the construction of ships. ·

The application of steam to ocean transport was one of the principal achievements of the nineteenth century. It is almost impossible to over-estimate the importance of the steam vessel in the development of commerce and civilisation. Regular services can now be carried on to all parts of the world independently of the influence of the weather. In the middle of the nineteenth century an ordinary voyage from Shanghai to London occupied 100 days; now the same distance is covered by the mail steamers in 35 days. Before the introduction of steam-ships on the American service, a voyage from New York to London lasted for thirteen or more days; now it is accomplished in five days or even less. It may here be mentioned as a matter of historical interest that the first steam vessels were successfully tried by Patrick Miller and others on Dalswinton Loch in Scotland in 1788, and in the following year on the Forth and Clyde Canal.

The other revolution was the introduction of iron instead

of wood as the material for constructing ships. In the first half of the nineteenth century English oak had become scarcer and more expensive, and foreign timbers were used with less success. Although iron was tried in constructing a small ship as early as 1787, it was not

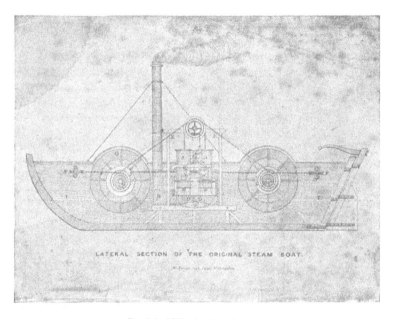

LATERAL SECTION OF THE ORIGINAL STEAM BOAT.

Patrick Miller's steamboat

successfully used on a large ship till 1821, and in 1837 the *Sirius* was the first iron vessel classed at "Lloyd's." Later mild steel was used instead of iron and has resulted in producing ships that are stronger, lighter, and more durable than vessels of either wood or iron. These steel-built vessels are subdivided into water-tight compartments, and thus the chances of lives being lost in a wreck or a collision have been reduced.

The introduction of iron and steel as the materials for construction has enabled sailing vessels to be built of

The *Aquitania*

much greater dimensions than in the olden days; and these large vessels are chiefly employed in carrying wheat and nitrate of soda from South America.

The early years of the twentieth century saw the introduction of the steam turbine, a wonderful invention due largely to Sir Charles Parsons. The first steamer fitted with turbines to cross the Atlantic was the *Virginian* in 1903. The great merit of the marine steam turbine is its compactness, and it is doubtful if machinery of the older type could have been got into such vessels as the *Aquitania* to develop the enormous 70,000 horse-power.

Further progress is being made in all departments of shipbuilding, and now coal is giving place to oil as a fuel. It is said that one ton of oil will produce as much power with a Diesel engine as three to four tons of coal with a steam engine. A remarkable feature of the shipbuilding industry has been the very great increase in the number of steamers for carrying petroleum in bulk, usually called "tankers." The largest tank steamers will carry about 15,000 tons of oil, and a great many of them are built to burn oil fuel.

4. BRITAIN'S MARITIME SUPREMACY

It has been well said that the people of our country have been made by shipping. Romans, Saxons, Normans, and other settlers all came here in ships and have made the British race what it is to-day. We are sea-born as a nation and our people are at home on the sea. Everyone realises the great part that shipping plays in our national life. We depend upon sea power for our security in these islands and for our foremost place in the world, and in a very real sense we depend upon our mercantile fleet and its protection by the Royal Navy for our daily bread.

An enormous amount of money is required for the building, equipment, and voyaging of all our ships, and as a consequence a vast number of men are employed. Indeed, after agriculture, shipping gives employment to the greatest number of people in the British Isles.

We have further to remember that the shipping of Britain is increasing. Other nations, such as Germany, have shown remarkable progress in the last ten years, but Britain has shown even greater progress and now has a tonnage exceeding by 50 per cent. that of the next six leading maritime nations combined. This maintenance of our maritime supremacy is of supreme importance to our Empire, for whereas formerly it contributed to our trade and safety, now it is indispensable to our very existence as a nation. The shipping trade is a most profitable source of revenue and the greater the trade of the world grows, the larger will be Britain's profits from her shipping.

Every child is taught that Britain is favourably situated so that it can trade with all parts of the world with much greater facility than any other nation. London, the capital of our Empire, and the first commercial and financial centre of the world, owes its premier position to our commercial supremacy at sea. London, in every sense, is the heart of the Empire and the connexion between the capital and the outlying members in the most remote regions of the world is a very real one. It is this commercial relationship that keeps us in contact with our far-flung empire; and that our merchant shipping is indispensable to us may be realised from the fact that we in Britain are dependent on other countries for three-fourths of our food supplies. The safety of our mercantile fleet is guaranteed by the British navy, so we have confidence that our food supplies from abroad will not fail to reach us.

The maintenance of British supremacy on the sea is due not only to the fact that we have more ships than other countries but also because those ships have better men. In Tudor times and later our British commanders were never afraid to fight with fleets greater than their own, and generally they won. What is true in our naval history is equally true in the history of our commerce. We have always had the best and the hardiest seamen to man our boats, and while that continues Britain need fear no rival on the sea.

It is almost impossible to exaggerate the importance of British shipping and shipbuilding industries: indeed if all the British merchant ships now afloat were lost at one stroke, the whole commerce of the world would be brought to a standstill.

Hence we not only want to know in a general way that the commerce of the British Empire is more extensive than that of any other state in the world, but we want to realise the immense importance of our fleets of trading ships and of the great part they play in the maintenance of our prosperity. As we have seen, the shipping industry ranks as the second largest of our national commercial pursuits, and we, as a nation, must do all we can to ensure that our boys and men are properly equipped to carry on the great traditions and the good name of our country.

We shall now understand the great importance of the Royal Navy, for its chief function is to keep the highway of the sea open for British commerce. Unlike the land the sea belongs to no one, for beyond a distance of three miles from the low-water limits of the shore, the sea is the common highway of all who sail on its waters. With the exception of its fisheries, which beyond the three mile limit are open to everyone, the sea is of no use except as a highway.   But its use as a highway is the most important factor in the economic life of a nation, which depends on sea-borne commerce for its supplies of food and of the raw materials for its industries.

Now Britain is such a nation, for no foreign merchandise can reach it or leave it except across the sea. As the sea is the sole highway for the foreign commerce of Britain, it is imperative to keep that highway open so that our ships may be secure in all their voyages.   The important point to remember is that our industries can be carried on and our people fed if the sea is kept securely open.   And this is the first and foremost function of the Royal Navy—to keep the seas open and to render our maritime commerce safe from attack.   In the official words the value of the Navy is thus stated: "It is on the Navy, under the Good Providence of God, that our Wealth, Prosperity, and Peace depend."

5. THE MERCANTILE MARINE, AND GREAT TRADE ROUTES

We have seen that the conditions of shipping have altered very much owing to the character of ship construction.   When wood was the material used in shipbuilding, all vessels were impelled by sails, and speed was governed by the variations of the wind and by the capacity of masts and sails to withstand pressure. When iron was introduced as the structural material and the steam engine became the driving power, both the size and speed of the vessels were increased. Towards the close of the last century further great advances in ship-building were made when steel replaced iron, and when the steam engine developed from a simple low-pressure machine driving a single screw into a high-pressure machine of double, or quadruple compounding. This latter, in turn, has been succeeded by the rotary steam turbine which is now used on the new vessels.

Now our mercantile marine may be said to embrace three classes of vessels. First, there is the ordinary cargo steamer, ready to go anywhere to pick up any sort of cargo, and with a speed ranging from eight to eleven knots. Second, there is the steamer suited for some particular line of trade, carrying a large amount of cargo in proportion to her tonnage, and with accommodation for passengers. Of this class are the coasting vessels round our coast and the majority of those trading from our country to America, to the Cape, India, and Australia. Third, there is the high-class liner, ever growing larger and more luxurious in fittings and accom-modation, and with a speed that has reached twenty-six knots.

These liners carry little cargo in comparison with their size and can only be profitably run where the passenger traffic is extensive and where there are ports that can accommodate them. While the largest and swiftest ships are mainly confined to the North Atlantic, there is a tendency to growth in size in cargo ships trading to the Far East, to South Africa, and to Australia.

The cargo steamer generally, and the tramp par-ticularly, is hired on charter by the merchant at a rate of freight which depends on the state of trade and the supply of shipping. Some steamers are chartered for a particular voyage only, while others are chartered for a period of time—six months, a year, or even a term of years. In the latter case the merchant who charters the vessel may send her anywhere, or may employ her only in certain latitudes or in a certain class of trade, and the owner of the ship loses direct control, but he has to stand the wear and tear.

Now let us consider the basis of a cargo steamer's work. Strange as it may sound, more coal is carried by tramps than by regular colliers, but it is carried in search of other cargo, and if freightage cannot be got at one port the steamers go off in ballast to some other port to get homeward cargo. The collier does nothing but carry coal round the coast, but the tramp only carries coal as one part of her work.

Many cargo steamers are built for special trades besides colliers. The grain carrier is a particular type, and the carrying of grain from Russia, from India and the Far East, from America, and from Australia is the most important service the cargo steamer does for us. Another branch of trade, for which steamers are specially constructed, is for the conveyance of cotton from America to feed the textile industry in which so many millions of our people are directly or indirectly engaged.

A modern form of the cargo steamer is the oil "tanker," for carrying petroleum in bulk, and the oil-tanker, after discharging the oil cargo, can be cleansed for holding a general cargo. Other lines of trade have specially constructed ships. Those that carry cattle from Ireland and America are liners rather than tramps, and are among the most useful of our merchant shipping. It is necessary that such ships should have speed, as is also the case in the carriers of frozen meat from North America, Argentina, and Australasia. In this latter class of ships, which are swift and well appointed, passengers are also carried.

The ships bringing butter and dairy produce from Holland and Denmark, and fruit from Spain and Portugal, come with rapidity, and discharge their cargo at whatever hour of the day or night they reach the docks. A steamer arriving at Hull or Newcastle at eight o'clock at night will have her cargo distributed in all the chief towns on the following morning.

Before the days of steamships, tea clippers used to come with great speed from China. Now tea is brought from China, India, and Ceylon by steamers specially constructed for the trade which make the voyage home without calling anywhere on the route. Speed, however, is now more important for butter, fruit, and meat ships than for tea, as the Suez Canal has shortened the voyage from the time of the old China clippers.

Some well-built tramps may make a series of voyages by taking coal or iron out to Brazil, then taking coffee from Brazil to New York, and afterwards grain or cotton for a third cargo to Britain. Sometimes, on the other hand, a boat going out to Australia with a general cargo may arrive there too late in the wool season or too early in the grain season. Then she has the choice to lay up till next season, or she may load a cargo of coal for some port in the Far East or on the Pacific Coast of America. Of all our British mercantile marine about 20 per cent. is entirely employed abroad and it is difficult to classify the trades in which such ships are engaged. About 70 per cent. of the cargo ships under the British flag may be classed as tramps, and it will thus be seen that the greater part of the world's carrying trade is

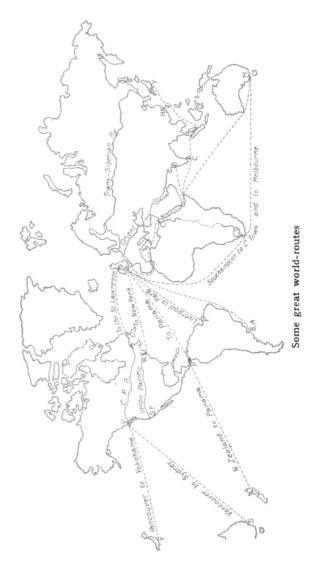

Some great world-routes

done by this useful class of vessel. It is also estimated that three-fourths of British exports, estimated in weight, are shipped by tramps either to foreign markets or to ocean coaling depôts.

Now it will be well just to consider which are the chief trade routes of the world on which cargo steamers are employed as well as cargo and passenger steamers. They are the North Atlantic trade; the North Sea and Baltic trade; the Mediterranean and Black Sea trade; the India, China, Japan, and Australasia trade; River Plate, West Indies, Brazil, Chile, and Peru, or South America trade; and the Pacific trade. These are the main routes, but of course there is some overlapping, and on them the tramps employed do about two-thirds of the total tonnage, while the remainder falls to the liners.

6. THE BUILDING OF A GREAT SHIP

Shipbuilding is one of the most important British industries and gives employment to nearly a quarter of a million persons. When ships were mainly built of wood, shipbuilding flourished in the neighbourhood of those ports which imported such woods as teak and pine, or where English oak and elm were accessible. In those days the Thames estuary was one of the chief centres of ship construction. With the change from wood to iron and steel it became necessary to have shipyards at ports near districts having large deposits of coal and iron, so that the cost of transit to the shipbuilding yards might be reduced to the lowest. For this reason the trade has left the Thames estuary, and the ports on the Tyne,

Wear, and Tees in England, and those on the Clyde estuary in Scotland, have become the most important centres of shipbuilding. In a recent year 973,000 tons of shipping were produced on the north-east coast of England, and 642,000 tons on the Clyde estuary. Other important shipbuilding districts are at Belfast, Barrow-in-Furness, on the Mersey, and on the Humber.

Let us consider briefly one of the great shipbuilding centres and for this purpose we cannot do better than turn our attention to the Clyde. From Glasgow to the mouth of the river the banks of the Clyde are lined with shipbuilding towns that make it one of the greatest shipbuilding centres of the world. We have already noted the immense tonnage of shipbuilding in one year, and it is obvious that the Clyde has many weighty advantages for this industry. The estuary runs into a busy coalfield where iron and steel working and marine engineering are leading industries, and thus material and skilled labour are close at hand. The enterprise and foresight of the citizens of Glasgow have changed a shallow stream into a navigable river, so that this industry may be possible. It is also noteworthy that some of the greatest inventions in shipbuilding were made by Clydesdale men.

After Glasgow the chief shipbuilding towns are Renfrew, Clydebank, Dumbarton, Port Glasgow, and Greenock. Renfrew has specialised in shipbuilding, for it builds more dredgers than any other town in the world. Clydebank has grown in a remarkable manner in recent years and has a number of large engineering and shipbuilding firms. Port Glasgow produces the largest tonnage of these shipbuilding towns, and Greenock is famous as the birthplace of James Watt.

Greenock, from the river

Now in order to make possible the large tonnage indicated by the figures already given, it is necessary that the shipyards shall be well equipped for all the purposes of constructing, launching, and completing vessels. A great modern shipyard must have convenient berths for building and launching ships, and it must also have a berth where a vessel may remain afloat after being launched and whilst being completed. It is very important that this berth should be close to the various shops concerned in the completion of the ship, so as to facilitate the work of the men going from the shops to the ships. This great shipyard should also have, or be within easy distance of, a dry dock, so that under surfaces may be painted and other bottom work done as the ship nears completion. It will thus be seen that the work in a shipyard has two phases. First, there is the building of the hull up to the point of launching, and secondly, the finishing of the ship after her launch and after the machinery has been placed on board. This machinery is not always put on board on the premises of the shipbuilder, but at the works of the machinery contractor. For this purpose a huge crane of 150 to 200 tons lifting power is required.

The ground on which a large vessel is built must be solid to withstand the great weight, and for very large vessels the berth is occasionally roofed in and sides provided to protect ship and workmen from the weather. In the middle line of each berth wood blocks are placed at intervals, and on these is placed the keel of the vessel about to be built.

Now before considering the workshops and the stages of building the ship a few words are necessary about the power plant. The machinery is usually electrically driven and the lighting of the works is through the same medium. The large shipyard tools have their own motors and the smaller ones are driven off a line of shafting. Much of the riveting is done by machinery, and the chipping of rough edges, hammering up of rivets in place, and caulking of plate edges by pneumatic tools. The capstans in a shipyard for hauling vessels in or out of dock are worked by electric or hydraulic power, as also are the cranes moving over berths. For a large shipyard the central electric power station may have an output of 6000 horse power.

Within easy access of the berths there must be many workshops. The plate and bar stores contain the steel intended for use, and cranes pick up the plates or bars as required and place them on the trucks or trolleys which run to the several shops. The mould loft is where the shape of the ship is determined and where information is given to the workmen as to the several parts of the structure. The bar furnaces heat the material intended for frames, and the plate shop has the plant for machining the hull plating. Here are punching machines for making holes in steel, shearing machines for cutting excess material from a plate or bar, drilling machines, and plate-bending machines. Among the other workshops may be mentioned the smiths' shop for forged bar work and forged fittings, and the shop containing the hydraulic and riveting plant. In the vicinity of the fitting-out berths there are saw-mills, plumbers' shops, paint shops, tinsmiths and galvanising shops, and many other minor but important sections.

It will at once be understood that to ensure the best results in a shipyard there must be organising skill to ensure harmonious working between the various departments.

For this purpose in the general office there is an estimating branch whose business is to deal with the cost and weights of ship work : a design branch which makes sketch designs for new ships; and a drawing office where all the detailed drawings are prepared. The head of the shipbuilding department is the general manager, and he deputes to the shipyard manager all duties relating to the actual building of the ship. The foremen receive their instructions from the shipyard manager and direct their workmen accordingly.

Let us suppose that a liner is to be built in a great shipyard. Before the contract for building such a ship is placed there is much preliminary work to be done. The design has to be worked out, and in the ordinary method wooden models of half the exterior surface and of the inner bottom are prepared, and on these are shown the main features of the new ship. Meanwhile some of the staff are engaged in the process of laying off, which consists in drawing the lines of the ship to full scale both in plan and elevation, so as to determine the dimensions of the chief parts of the structure. It is now probable that the materials are being obtained and everything is ready for the building of the ship to be begun.

The keel is first placed in position and extends from one end of the vessel to the other on the building blocks. Then follows the framing which is designed to stiffen the bottom, and side plating to resist water and other pressures. The way is now prepared to fix the shell plating whose function is to keep water from the interior of the vessel, and when this work has sufficiently progressed, a beginning is made with the several steel decks, leaving proper spaces for the machinery which is shipped after the vessel is launched. Thus day by day, week by week, and month by month, the work goes on, and what in the early stages was merely the skeleton of the ship takes shape and form.

When the ship is ready to go into the water, all the scaffolding is removed, the building blocks and shores are knocked away, and the huge vessel is held by a few wedges and a cord. The final act of launching, often performed by a lady, consists in cutting a string, and then turning a handle to release weights, so as to leave the vessel free to move down the ways into the water.

After the launch the ship is towed to a suitable crane berth to get her machinery on board, and then moored to a position where the work of completion is taken in hand. Steel decks are secured; wood decks are laid, necessary cabin bulk-heads are erected, and cabins fitted out; heating and lighting arrangements installed; and all the other necessary details are completed.

The vessel is now ready for the trial trip, so that the speed may be determined, and the consumption of fuel to reach that speed. The trial is either run between given ports at the required speed, or for a fixed number of hours at the given speed. If everything is satisfactory at this trial, the ship is fit for sea, and with the captain in command and the crew on board, she is readv for her first voyage.

7. THE ROYAL NAVY

The Royal Navy owes its origin to Alfred the Great, who founded a fleet and kept it in a thoroughly efficient condition so that England might be safe from invasion by the Danes. From the ninth century to the present time, for more than a thousand years, it has been apparent that England must look to the Navy for her first line of defence. There have been periods in our history when our rulers have been indifferent to the strength and efficiency of the Navy, but these times of weakness soon caused the nation to awake and make good the deficiencies.

It was in the time of the Tudors that our Navy was refounded, and the work of Henry VIII was most encouraging as he realised the necessity of having good ships, skilful sailors and proper dockyards. The defeat of the Armada in 1588 and our wars with the Dutch in Stewart times were factors in giving our seamen a chance to show their mettle, and for a long time no foreign power seriously attempted to break through the line of the British fleet.

Then came a dark time in the eighteenth century when our Navy declined once more, and, as one of the results, our American colonies were lost. Englishmen, however, rallied to the support of the Navy and under the splendid leadership of Hawke, Rodney, Howe, and Nelson, it was able to assert its supremacy which culminated at Trafalgar in 1805. From that memorable year to the present time England has been mistress of the seas, and until the Great War of 1914 no nation had challenged that supremacy.

In former times the ships of the Navy were built of wood—"heart of oak are our ships"—and it was not till the first half of the nineteenth century that a great change came in the Royal Navy with the introduction of steam-power, which, in 1832, was fitted to the *Salamander*. The results were so successful that each subsequent year saw steam-power applied to all types of ships in the Royal Navy. Another remarkable change in the ships of the Navy was made when ironclads were built, for an end was put to wooden ships by the introduction of shells about the time of the Crimean War. The first iron-built war vessels were built in 1856, and the first English ironclad, the *Warrior*, was launched from the Thames in 1860. That vessel marks a great advance in the construction of warships, which are being constantly improved in every respect as each new type is evolved.

When our ships were built of wood, the largest vessels were called Line-of-Battle-ships. They were classed as first, second, or third-raters, according to the weight and number of guns carried. Below this class the other ships were ranked as Frigates, Sloops, and Corvettes. The *Victory*, which is now in Portsmouth Harbour, is a good type of an old wooden three-decker, and when we compare Nelson's flag-ship with the monster steel battleship of to-day we realise how great is the change made in the ships of the Royal Navy in less than a century.

At the present time the ships of the Royal Navy may be classed as armoured and unarmoured. Armoured ships are those whose sides and guns are protected by plates of armour; while unarmoured ships are those without protective armour. The largest of the armoured ships are Battleships, which are required to blockade an enemy's ports, to bombard fortified places, and to engage other armoured ships. Nelson's ship, the *Victory*, carried 104 guns, and the weight of the broadside of her 52 guns was 1160 lb. Now compare this with the modern battleship and we find that the weight of one shell from the 13·5 inch gun is 1400 lb.

The Cruisers are smaller than battleships and are of various classes according to size, number of guns, weight of armour, and age. They are used for patrolling the seas, guarding merchant ships, preying upon the enemy's fleet, or for scouting. Like the battleships the armoured cruisers have been built in batches, and with their high speed and formidable armament they are most efficient vessels.

Destroyers rank next in importance. They are of very high speed, and well qualified to run down the torpedo craft of an enemy's fleet, and to perform scout duty. Destroyers are unarmoured, carry only the lightest guns, and are usually fitted with some torpedo-tubes.

The submarines are among the latest developments of

Submarines at Dundee

our Navy and came to the front at the latter end of the nineteenth century. They are used for the defence of our harbours and water-ways, and, as they can be made to sink below the surface and discharge torpedoes at an enemy's vessels within range, they are of a very destructive character.

Besides the types of ships already mentioned there are others of a miscellaneous nature. There are torpedo depôt ships that attend to the needs of torpedo-boat and destroyer flotillas; and submarine depôt ships that perform the same offices for the submarines. The mine-laying squadron is kept ready to mine any area that it may be wise to defend: and the mine-sweepers are old torpedo boats and steam trawlers that have been fitted out for this purpose.

The Royal Navy has several ships employed in the surveying service, for use as his Majesty's royal yachts, coastguard cruisers, troopships, and river gunboats. Nor must we omit to mention the merchant cruisers, which include the largest and fastest of the great liners belonging to the leading steamship companies. These merchant cruisers are all of high speed, have great coal-carrying capacity, and can be armed with a number of light guns.

Among the recent developments of the Royal Navy is the formation of the Royal Naval Air Service, which has the control of all air-ships and sea-planes and possesses a number of overland flying machines. Sea-planes or hydroplanes are at present in an undeveloped stage, though Britain has made more progress in their use than any other country.

When the Navy and the Army are employed together the Navy takes the right of the line or heads the column on the march. It has this privilege because it is the senior service, there having been a disciplined navy before there was a standing army. The head of the Royal Navy is his Majesty King George V, who is a seaman in fact as well as in name, for he has served in the junior ranks, and commanded a torpedo boat, a gunboat, and large vessels of the Fleet. At the end of 1914 there were about 600 ships in the Royal Navy, and to man all these ships there were 152,000 officers, seamen, boys and others.

8. THE ROYAL DOCKYARDS

The Royal Dockyards are maintained for the purpose of building, repairing, and equipping with men and stores the ships belonging to the Royal Navy. Before the reign of Henry VIII there were no dockyards for the King's ships, and so that Tudor monarch, when he re-founded the Navy, formed the first dockyard at Woolwich. Since that time later monarchs have established dockyards at Portsmouth, Deptford, Chatham, Sheerness, Devonport, and Pembroke. Of those mentioned Woolwich and Dept-ford have long since ceased to be dockyards, but all the others are kept in a high state of efficiency.

Speaking generally these dockyards have slips for building ships, workshops for various manufactures, dry docks for repairing, and stores for keeping arms, ammuni-tion, coal, and provisions. Some of the dockyards no longer build ships and are little more than harbours where a ship can replenish its stores and carry out minor repairs. Many of the great private firms on the Tees, Tyne, and Clyde, at Barrow, Belfast, and elsewhere are engaged in the construction of the ships to an advanced stage, and then they are sent to the Royal Dockyards to be completed and equipped.

Before describing the Royal Dockyards it may be well to explain that the chief officer in charge is known as the superintendent. In the largest establishments he is a rear-admiral and in the smaller yards a captain. Under him are several heads of the various departments who have charge of thousands of workmen. The actual

work of building the ships in a Royal Dockyard is in
charge of the Chief Constructor who acts in accordance
with plans from the Admiralty.

Portsmouth, the largest naval establishment in the
world and the strongest fortified place in Britain, must
first claim our attention. It is the aggregate of four
towns—Portsmouth, the garrison town, Portsea, the site

Portsmouth Harbour and Dockyard

of the great naval dockyard, Landport, the residence
of the artisans, and Southsea, a modern watering place.
Portsmouth has an admirable situation, placed as it is
on the south-west of Portsea Island and between Portsmouth Harbour and Langstone Harbour. There is communication by a ferry and floating bridge with Gosport
on the opposite coast where are the victualling yard and
Haslar Hospital.

Portsmouth was established as a royal dockyard in 1540, though before that date it had been a naval station for the King's ships. Cromwell increased its importance, and Charles II still further extended its capacity. Since then at various dates it has been enlarged, so that now the dockyard covers more than 300 acres of basins, docks, slips, factories, workshops and storehouses. There are fifteen dry docks, nearly 20,000 feet of wharfage, and ten miles of railway. The harbour is constantly being dredged, so as to increase the berthage accommodation for the ships of the fleet.

There is a gunnery establishment at Whale Island and torpedo ranges in the harbour. The extensive coal wharfs have hydraulic appliances for supplying ships with coal, and the chief shore depôt for submarines is at Portsmouth.

The enclosed basins are used for ships that need repairs above-water, and around these extensive ponds are factories and workshops, and shears and cranes which are worked by steam power.

The dry docks are used for building ships or for repairing the under-water parts of ships. When the ship enters the dock, the entrance is closed by the dock gate, and then the water is pumped out. The ship is now in a large dry basin, with her keel resting on a long line of iron blocks, while she is kept upright by stout supports to the sides.

Portsmouth Dockyard has iron and brass foundries, machine, boiler, and coppersmiths' shops, and smithies. In the largest smithy there are 120 fires and as many as twelve Nasmyth steam-hammers, one of which weighs seven tons. There are also rope, mast, block, and boat-making houses, and storehouses for torpedoes, chains, and cables. Specially interesting are the wood-mills which contain ingenious block-making machines by which about 60,000 blocks are made annually.

Portsmouth Harbour is a spacious body of water extending four miles inland and having an extreme width of three miles. The entrance to it is 400 yards across and the largest vessels can gain access at any time.

Devonport, one of the "three towns"—Plymouth, Devonport and Stonehouse—was established as a dockyard by William III in 1689 when it was known as Plymouth Rock. As a naval station Plymouth is second only to Portsmouth, the spaciousness of the Sound affording a fine anchorage to a large number of ships. The Port of Plymouth includes all the water of Plymouth Sound and the Hamoaze, including all bays, creeks, lakes, pools, ponds, and rivers as far as the tide flows within and northward of a line drawn across the entrance of Plymouth Sound. The chief waters are the Catwater, Sutton Pool, Mill Bay, Stonehouse Pool, and the Hamoaze.

Plymouth Sound has an extent of 4500 acres and is sheltered from south-west gales by a breakwater $2\frac{1}{2}$ miles south of the Hoe. It is nearly a mile long and was constructed in 1841 at a cost of £1,580,000. Plymouth is well fortified; there are forts on the Catwater and Hamoaze commanding the harbour and its approaches. On the Hoe there is Smeaton's tower, now used as a wind-recording station, Drake's statue, the Armada Memorial, a Bowling Green which reminds one of history. In the Sound is Drake's Island which is strongly fortified.

The dockyard, arsenal, and other government establishments lie along the Hamoaze and were considerably extended in 1907.

The dockyard has three features. Its tidal basin is

740 feet long: the three graving docks with entrance locks are 745 feet long; and the large enclosed basin with a coaling depôt is 1550 feet long and covers 35½ acres. The direct entrance to Hamoaze is closed by caissons. Compressed air is used to work the sliding caissons, which close the entrance of the locks and the closed basins. It may be mentioned that Devonport supplies half the hempen rope used in the Royal Navy.

Chatham Dockyard

Chatham is one of the chief naval arsenals of Britain, and the first dockyard here was established in 1588. Since that date it has been much extended and now it covers an area of 516 acres and has a river frontage of three miles. Before 1867 there was no enclosed basin or wet dock; but since then a portion of salt marshes has been taken in, and now there are several tidal docks, building slips and other docks, capable of turning out the largest men-of-war.

Chatham is one of the chief defences of London and is strongly fortified. The fortifications are very elaborate and include, besides forts, an intricate system of trenches, batteries, and subterranean passages. It is a matter of history that the Dutch under De Ruyter, in 1667, sailed up the Medway as far as Chatham, but did little damage.

Sheerness is a dockyard chiefly of use for naval repairs. The harbour is safe and commodious, with deep water close alongside the quays at low tide. The dockyard was commenced in 1814, and covers an area of 60 acres. There are three basins and five docks, besides extensive naval stores and barracks for sailors and soldiers. Charles II built a small fort here after De Ruyter's visit on 10th July, 1667.

Pembroke, on a navigable creek of Milford Haven, has a dockyard which covers ninety acres and is enclosed by high walls and strongly fortified. Ironclads are constructed and repaired, and at Pennar Gut there is a sub-depôt with accommodation for artillery and infantry. Pembroke Dockyard is the latest of the dockyards and was not commenced till 1814.

9. THE FISHING INDUSTRY—SHIPS AND MEN

We can quite easily understand that there should be extensive fisheries around the coasts of the British Isles, but it is not always remembered that these fisheries employ more than 100,000 men and boys, and that the annual value of the fish landed on our shores is estimated at

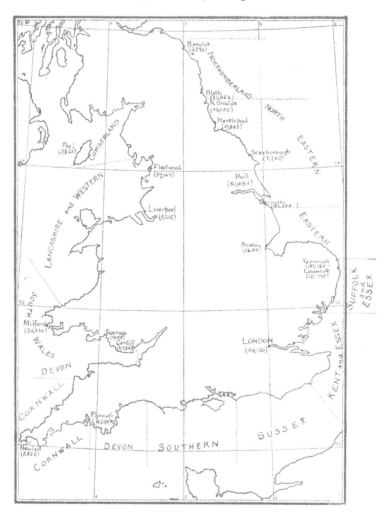

Map of England and Wales showing the Sea Fisheries Districts and the Chief Fishing Ports. The figures attached to each port show the landings in tons of fish caught during 1913

nearly £12,000,000. During the last twenty or thirty years mechanical enterprise has completely changed the British fishing industries, and it has also made the lot of the fisher more tolerable than it was in the olden time.

The dweller in an inland county invariably thinks of the fishing being undertaken by sailing fleets, and certainly there is something picturesque in the idea of these smacks with their brown or red sails setting out in the evening and letting down their nets for a draught. It is true that there are still fishing smacks at such old world ports as Ramsgate and Brixham, but the main work of fishing is by means of steam trawlers.

One of the first steam trawlers was built at Hull in 1881, and so successful were the early steam trawlers that in a very few years the sailing fleets were nearly all superseded. These steam trawlers are among the finest little sea-boats in the world. They are splendidly equipped and well found in every way; indeed they are the result of long years of experience. Their construction has brought into existence large industries in towns like Selby, Beverley, North Shields and Aberdeen. Some of the finest trawlers in the world are built on the Ouse in the little market town of Selby, right in the heart of Yorkshire. The fishing vessels there built and equipped are of the largest size, and so big are many of them that, owing to the narrowness of the river, they are launched broadside on.

A vessel of this type, constructed expressly for fishing, is about 140 feet long with a gross tonnage of 300. The speed of the ship is from ten to eleven knots and the cost not less than £7500. It has been well remarked that these trawlers are to the fishing world what the great liners are to the mercantile marine. After the steam trawlers come the steam drifters and various kinds of motor-driven vessels.

Around the coasts of Great Britain the number of first-class steam trawlers is probably not far short of 2000. In England and Wales the trawling industry is by far the most important form of sea-fishing. In Scotland it provides employment for less than ten per cent. of the fishermen, while in England it gives work to more hands than all the other methods of fishing taken together.

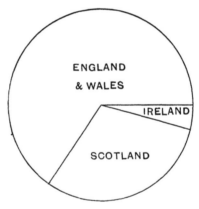

Diagram showing comparative quantities of Fish landed in England and Wales, Scotland and Ireland in 1913

About thirty years ago trawling was mainly confined to the North Sea, and the work was done by sailing vessels that were entirely dependent on the weather. At the present time there are eighteen "regions," or fishing grounds, frequented by British steam trawlers. The foreign grounds are the White Sea, the coast of Norway, the Baltic Sea, Iceland, Faröe, the west of France, the north of Spain, the coast of Portugal, and the coast of Morocco. The home grounds are the North Sea, the

north of Scotland, the west of Scotland, Rockall, the west
of Ireland, the Irish Sea, the south of Ireland, the Bristol
Channel, and the English Channel. The largest of all
these areas is the North Sea, which is regularly worked
by four fleets and a very large number of single-boaters.
The old fleeting system continues, the vessels forming the
fleets remaining on the grounds under the control of
their admirals, and sending their fish to market by steam
carriers. More fish are brought every year into Grimsby,
Hull, Lowestoft, and Yarmouth than into all the other
fishing ports of England put together.

While trawling, or the dragging of a large net across
the bed of the ocean, is the most important branch of the
fishing industry, the herring fishery is scarcely less
important. At the proper season of the year vast fleets
of craft, both steam and sail, are engaged, and the
enormous number of these ships make Yarmouth,
Lowestoft, and Grimsby very busy during the autumn
months. Not the least interesting feature of these fleets
is their varied nature, for they include vessels from the
small coble manned by three men to the imposing steam-
drifter, which is the largest type of vessel employed in
catching herrings. The boats engaged in the herring
fishery are generally classed under the title of "herring
drifters," but their individual titles vary according to
the locality they hail from and the type they belong to,
and include such names as sploshers, cobles, yoffers,
keel-boats, luggers, yawls, and mules.

10. BRITISH FISHERIES AND FISHING PORTS

There are many species of British sea-fish and of these
about fifty have a marketable value, although only
about thirty are caught in considerable quantities.
They are divided into two classes. Some are called
demersal, because they live and feed near the bottom of
the sea; the others are called *pelagic*, because they swim
about in shoals near the surface of the sea. Of the
former class the sole and the plaice are examples, while
the herring and the mackerel will represent the latter class.

Now it will be evident that different methods must be
used for capturing these two classes of fish. The former
are chiefly taken by trawling vessels, which drag their
nets along the bottom of the sea, and the latter are taken
by "drifters," which spread their nets just beneath the
surface of the sea. Besides these methods, a small
quantity of fish is taken by "liners," that is, vessels
fishing by means of lines of baited hooks.

It has been already mentioned that the principal
British fishing grounds are in the North Sea, and we may
now add that about six times as many demersal fish and
about fifteen times as many pelagic fish are landed every
year at east coast ports as at ports on the west coast.

Grimsby, Aberdeen, and Dublin are the headquarters
of the English, Scottish, and Irish sea fisheries. The two
first-named ports give remarkable evidence of the swift
advance of steam trawling enterprise. In 1889, Grimsby,
which was then, as it is now, the largest fishing port in

the world, had a fleet of more than 800 vessels of which only thirty-eight were steamers. At the present time there are 700 steam fishing vessels belonging to Grimsby, and the crews employed number 6000 men and boys. In 1854 Grimsby despatched 453 tons of fish, whereas in a recent year the total had risen to 179,792 tons. Hull also has an enormous interest in North Sea fishing and has no less than 436 first-class steam trawlers.

Now let us turn to Aberdeen, whose growth as a fishing centre has been most remarkable. In 1891 only fifty-nine steam trawlers and six liners belonged to this port; now the trawlers number 217 and the liners fifty-three. The Aberdeen fish-market is a successful municipal enterprise and yields a revenue of nearly £20,000 a year.

The Scottish fisheries have been revolutionised by the introduction of mechanically propelled vessels. The auxiliary motor-boats on the north-east coast occupy a position between steam-drifters and sail-drifters, and the enterprises of these motor-boats have met with much success. Thus a small motor skiff in Shetland earned £607 in a recent year, and of that amount probably seven-eighths were net profits.

The herring fisheries are closely connected with Yarmouth, the principal fishing port in Norfolk, and with Lowestoft, the chief fishing port in Suffolk. Yarmouth formerly carried on an important trawl fishery, and its sailing trawlers in 1889 numbered 400 vessels. The introduction of steam trawlers at Grimsby and Hull helped to ruin this industry. To compensate for this loss the Yarmouth herring fishery has increased enormously, and in October and November, when the autumn fishing is at its height, the busy scene is striking and interesting. The Yarmouth herring fishery is carried on by both sailing and steam fishing boats or "drifters." The Yarmouth boats are distinguished from those of other ports by having YH painted in large white letters on their sides. During the autumn fishing the Yarmouth fleet is joined by a much larger fleet of Scottish fishing boats, usually numbering about 600. Herrings are caught all the year round off some part of the British coasts; but at Yarmouth the chief season is from September till the middle of

Yarmouth steam drifters

December, when the herring shoals arrive off the Norfolk coast. The fishing is at night, when perhaps 2000 miles of nets are spread on the fishing grounds.

Lowestoft carries on an important trawl fishery, its fleet of steam and sailing trawlers being one of the finest in the world. Its trawlers and drifters comprise 529 vessels of which 158 are steamers. These vessels are distinguished by the letters LT on their sides.

It is worth noting that of all the herrings landed at the thirteen principal herring-fishing ports, one-third of the total was landed at Yarmouth and one-fourth at Lowestoft. The fishing industry finds employment for many thousands of men at these two East Anglian ports. Probably there are 7000 or 8000 fishermen living in Yarmouth and Lowestoft, besides many hundreds in the neighbouring villages. In addition to these numbers about 7000 Scottish fishermen have their headquarters at these two ports in the autumn, and several thousand Scottish girls come to work in connexion with the kippering of herrings. A very large proportion of the herrings caught by the Yarmouth and Lowestoft fleets is exported to Germany, Russia, and Italy.

The fish of the English Channel differ considerably from those of the North Sea. Haddock, one of the most abundant species on the east coast, is very rare in the south. The cod, again, is a northern species, and is almost entirely absent from the Bristol and English Channels. Whiting is one of the most abundant of English Channel fish; and in this species, as well as in soles and turbot, the south coast is second only to the east coast. Conger-eels are abundant in the English Channel, and there are plentiful supplies of gurnards, skate, and dogfish. Pilchards are confined to Cornwall and to the south coast of Devon, but by far the larger quantity is taken off the former county. Newlyn, Plymouth, Brixham, Torquay, and Newquay are the most important fishing stations on the south-west coast, while Fleetwood and Milford take first and second places respectively on the west coast.

Besides the various fish already mentioned as being caught around our coasts there are also many kinds of shell-fish, such as oysters, lobsters, crabs, cockles, and whelks, which give employment to hundreds of fishermen.

Essex oysters are deservedly famous, and those taken
from the beds at Burnham in Essex and in the Colne
fetch a high price.   Some of the largest English oyster
beds lie off Whitstable in Kent and the "natives" from
that town have a great reputation.   Cromer has for a long
time been famous for its crab fishing, and Sheringham
has even more boats employed in this fishery than
Cromer.   Lobsters are also caught in large numbers at
these places.   In Suffolk both lobsters and crabs are

Pilchard boats

chiefly obtained off Thorpe and Aldeburgh.   The season
for crab and lobster fishing is from the middle of March
till the second week in October.

  With regard to the Irish fisheries their development
has not been so marked as in the other fishery centres of
the British Isles, but there has been some progress both
in the quantities caught and their cash value.   The Irish
fishing vessels are generally poor and there is a remarkable
absence of mechanical appliances.   The herring fisheries

off the Irish coast show most improvement, but this is largely due to the enterprise of English and Scottish fishermen.

There have been many changes during recent years in the system of fishing around the British coasts; and it is likely that in the near future there will be even greater developments. It is probable that wireless telegraphy will be installed in the fleets, and experiments now being made with turbines and oil engines may tend to lessen the cost of the upkeep of the fishing vessels.

Some reference may be made to the more important by-products of the fishing industry and to the efforts that are being made to use the valuable material which might otherwise go to waste. The development of the manufacture of fish meals and manures progresses steadily, and during the last few years some of the large steam trawlers have been fitted with plant for the manufacture on board of manure and for the extraction of oil. These trawlers are thus able to deal with their entire catch in such a way that no part of it possessing any economic value need be thrown overboard. The value of fish oil for medicinal and other purposes has long been recognised, and there is every reason to suppose that its value may increase in the near future.

With regard to our foreign trade in fish it is worth noting that about 11,000,000 cwts. of fish were exported from the United Kingdom in 1913. The herring plays by far the larger part in this export business, and next in importance to the herring comes the cod which is chiefly exported in a dried form. There is no doubt that the expansion in our fishing enterprise of recent years has been owing to the increased continental demand for fish and fishery products.

11. "LLOYD'S." THE PLIMSOLL MARK. A1

"Lloyd's," which is short for Lloyd's Subscription Rooms, is now an association of underwriters whose business is largely to do with shipping and insurance. This great Maritime Exchange is really part of the Royal Exchange, and owes its origin to Edward Lloyd who kept a coffee house in Tower Street in the seventeenth century. As was the custom in those days, Lloyd's Coffee House was a place where news could be obtained; and to-day "Lloyd's" is the leading institution in the world for obtaining maritime news.   In one of the rooms may be seen the first Insurance Policy of which there is any record.   It was taken for a ship, the *Golden Fleece*, which was insured for a voyage from Lisbon to Venice for £1200 at four per cent. on January 20th, 1600.

Marine Insurance is the main duty of underwriters at Lloyd's, and for this business the members and subscribers to Lloyd's are classified as the underwriters, or those who accept the risks, and the brokers, or those who place the insurance on behalf of the owners of the ships or cargoes.

An underwriter must possess a minute knowledge of ships and their owners, as well as of the conditions of the trade in which they are engaged.   For this purpose the members of Lloyd's have reports from their correspondents all over the world, and are thus enabled to know the condition of almost every ship afloat.   A great deal of this information is published daily in the official *Lloyd's List*, which is the second oldest newspaper in Europe. Extracts from this are also published in most of the daily papers, so that we are enabled to know about the sailing

of ships, the ports at which they touch, and any accidents
they may incur.

Then the underwriters have another source of information known as the *Captain's Register*, in which is
recorded the chief facts in the seafaring life of every
captain, the names of the ships he has commanded, the
accidents or disasters connected with his management of
those ships, and the creditable and heroic services he may
have rendered. All this information about the ships
and their captains is of the utmost importance as it
affects the mind of the underwriter when he is considering
the question of insurance.

Here it may be mentioned that whenever a ship has
been wrecked or met with a serious accident, the whole
circumstances are entered in Lloyd's *Loss Book* which is
posted up from day to day. In the Underwriters' Room
there is a famous bell taken from H.M.S. *Lutine*, which
sank in the Zuider Zee in 1799 carrying down with her
treasure to upwards of one million sterling. This bell is
tolled once upon the announcement of a ship's loss and
twice when there is news of an overdue ship. A ship is
never posted at Lloyd's until all hope is gone and then all
insurances on the lost ship become payable, and the crew
and officers are considered dead in the eyes of the law.

There is also an enquiry office at Lloyd's where the
relatives of the passengers or crew may obtain free any
information concerning the vessel in which they are
interested. The Corporation of Lloyd's present medals
to those who have helped to save life at sea, and also to
officers and others who have contributed to the preser-
vation of their vessels or cargoes when in great danger.

It has been already mentioned that Lloyd's agents
are in all parts of the world, and their duties are to render

advice or assistance to masters of vessels in case of ship-
wreck; to report by telegraph direct to Lloyd's all
casualties within their district; and generally to report,
under guiding rules, all marine happenings within their
knowledge.

Lloyd's signal stations are in all parts of the United
Kingdom and abroad, and are of the utmost value to the
shipping world. If a shipowner or some interested party
wishes to communicate with any vessel at any Lloyd's
signal station, he has only to signify his wish to the head
office and instructions will be given accordingly.

In connexion with this subject it will be well to
devote one or two paragraphs to the Load Line Marks.
Not very long ago there was an idea that ships were sent
to sea so defective or overloaded as to render it almost
impossible for them to reach their destination in the
event of rough seas. It was largely owing to the agitation
by Mr Plimsoll in 1873 that improvements were made,
and now all British ships, with few exceptions, must be
marked with load-lines in accordance with certain tables.
These marks, generally called Plimsoll marks, consist of
a disc with a horizontal line running through the centre
and extending a little on each side of the circle. In
addition to this there are on steamships a number of
lines at right angles to a perpendicular, which indicate the
load limit for different seasons of the year and for fresh
water.

A reference to the diagrams will illustrate the meaning
of the Plimsoll Mark. The marking I.S. means Indian
Summer; S. Summer; W. Winter; and W.N.A. Winter,
North Atlantic. The F.W. signifies Fresh Water and
represents a provision for deeper immersion in fresh
water. The lines show the maximum depth to which a

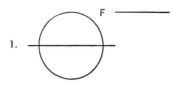

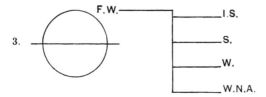

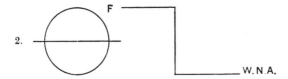

1.   Markings on the Starboard side of a sailing
     ship engaged in coasting trade.
2.   Markings on the Starboard side of a sea-
     going sailing ship.
3.   Markings on the Starboard side of a
     steamship.

     F.W.   Fresh Water.

     I.S.   Indian Summer.

     S.   Summer.

     W.   Winter.

     W.N.A.   Winter North Atlantic

The Plimsoll Marks

vessel may be loaded for a voyage in question. Thus for a summer voyage the S. line will appear on or immediately above the surface of the water.

All sound sea-going ships are classified by Lloyd's Register[1] as A1, and retain their characters as long as they are found to be in a fit and efficient condition to carry their cargoes. Every ship must be specially surveyed every four years, or oftener if necessary. It is generally understood that A1 means first-rate—the very best. The character of the ship's hull is designated by letters, and that of the anchors, cables, and stores by figures. A1 thus means hull first-rate and also anchors, cables, and stores; A2, hull first-rate, but furniture second-rate.

12. TRINITY HOUSE AND ITS WORK

Trinity House is the general way of speaking about the Association of English Mariners which dates from the reign of Henry VIII, and is now known as the Corporation of Trinity House, having its headquarters at Tower Hill, London. At the present time there are thirteen Elder Brethren, two from the Royal Navy and eleven from the Mercantile Marine. The work done by the Elder Brethren of Trinity House is of the utmost importance, for it is "the duty of erecting and maintaining lighthouses and other marks and signs of the sea."

The income of Trinity House is derived from light dues levied on shipping, and by this means the lighthouses, lightships, beacons, buoys, etc. are erected and maintained

---

[1] "Lloyd's" must be distinguished from Lloyd's Register which was a Society formed in Fenchurch Street in 1843. It publishes annually an accurate Register of all ships using British ports.

round our English coast. What Trinity House does for us is to make sure that the ships may approach and enter our harbours with safety. When we remember that the greater part of our food supply comes from abroad over the sea, we shall then understand how vital it is that it should arrive with regularity and punctuality.

The dangers of our coast to the mariner are very numerous. To mention only a few, there are such

Trinity House

outstanding rocks as the Wolf, the Bishop, and the Eddystone which are far from land and in deep water; and there are shoals and sandbanks at the entrance to the Thames, along the Essex coast, and elsewhere. Trinity House has the task of seeing that the coast is properly lighted, that pilots who know the difficulties of the coast are appointed, and that all wrecks shall be removed that are dangerous to navigation. And here it is worth noting that whereas 100 years ago there were only about thirty lighthouses and lightships around the British coast, there are now about 900.

Once every month the Trinity House steamer visits the lightships and lighthouses that are actually detached from the land. The Trinity House steamer may easily be recognised in our harbours as it is painted in the colours that once distinguished our battleships; and its duty is to visit and inspect all light vessels, supervise repairs, tow them out and home when required, and generally to look after all the coastal arrangements made for the safety of mariners.

When the water is smooth the monthly reliefs by the Trinity steamer are easy, but the relief of the rock light-houses in the open sea is both hazardous and difficult, for the men are swung out of boats in baskets and thus raised to the tower. Of course there are accidents, but these are comparatively few owing to the skill and care of the men who do this work. The faithful and efficient service of all concerned with Trinity House, from the Elder Brethren to the lighthouse-keepers, deserves the warm gratitude of all people in our country and of those who visit our shores. "The coast-wise lights of England watch the ships of England go," and that these lights may shine the mariner trusts to Trinity House.

Besides lighting the coasts most countries have some uniform system of buoyage. There are a few types of buoys that are so peculiar as to deserve notice. The gas buoy has its recognised number of flashes per minute in order that it may be at once recognised, and has an ingenious arrangement, in the nature of a clock, which turns off the gas at sunrise and turns it on again at night. The gas is pumped from the Trinity steamer into the interior of the buoy, the supply lasting from one to two months.

Then there is the curious whistling buoy which has not proved very successful. The whistle is fixed to the top of an ordinary buoy and its doleful sound may be heard on some of our waters. The mode of working of this buoy is as follows. The whistle is connected by an iron pipe descending into the water through the middle of the buoy. As the buoy rises and sinks, the water in this tube rises and falls, and so compresses the air, which is forced through the whistle, thus causing the sound, which, however, is not very loud.

The most familiar buoy is the bell buoy, and in this type a bell is firmly fixed inside an iron cage mounted upon the hull of the buoy. Attached to the cage are some clappers which violently hammer the bell as the buoy rolls and rocks.

The variety of buoys is endless but they all have distinctive meanings. Thus a green buoy denotes that there is a wreck in its vicinity; while red buoys are on the starboard side when entering from the sea, and black buoys on the port side. Watch-buoys are of can-shape and are sometimes placed near a lightship. Their colour is red and they bear the word WATCH, preceded by the name of the light-vessel in large white letters. It will thus be evident that buoys are important guides in navigation, and by their many shapes and colours are of peculiar significance to the sailor.

13. LIGHTHOUSES AND LIGHTSHIPS

Lighthouses or beacons for the guidance of seamen have a long history. They date back as far as B.C. 331, when the famous Pharos of Alexandria was in position; and in our own land there is the celebrated Pharos at Dover which was built when the Romans lived in Britain. During the Middle Ages some of the prominent headlands were lighted by the charity of religious houses, as, for instance, St Alban's Head in Dorset.

In its earliest form the lighthouse was simply a tower or beacon, on the top of which wood or coal was burned in a brazier or iron basket; and this arrangement on isolated rocks or promontories continued with very slight modifications or improvements until the nineteenth century. Smeaton's celebrated Eddystone Lighthouse of 1759 was lighted by twenty-four candles, and later on the illuminating power of a lighthouse was improved by the use of mineral oils. In more modern times there have been great improvements by the invention and development of new appliances for the burning of mineral and animal oil. Gas has also been used with good results, and the electric light is used in some well-known lighthouses, especially at St Catherine's Point and Cape Grisnez. The electric light, however, has not proved so successful as gas in a foggy atmosphere.

In the matter of lenses and reflectors, the fittings of lighthouses have been much improved, so as to avoid loss of light and extend its power. To concentrate the rays of the lamp and disperse them in the right direction,

Thomas Stevenson in 1849 employed a lens in front of a reflector, which combined both the direct and the reflected rays into one parallel beam. Other improvements have followed and now various French and English systems are in use at the lighthouses around our coasts.

Lighthouse construction varies with the demands of the locality and the form of the building with surrounding circumstances. The building of a lighthouse such as the South Foreland, on the top of a cliff beyond the reach of waves, is a much simpler problem than one on an isolated rock, as in the case of the Eddystone, or of the Bishop's Rock in the Scilly Islands. There the foundation of the lighthouse is quarried out of the solid wave-swept rock, and must be so strong as to withstand the Atlantic breakers and the fiercest hurricane. The solid foundation of the tower is carried a score or so of feet above the surface of the water, when it is surmounted by a lofty hollow tower in which are various rooms for the keepers, and in this form it is continued to the top, in which the lantern is placed.

When lighthouses are placed in an estuary, or upon a spit or tongue of sand, piles have to be driven, and the lighthouse, broad based and made of wrought iron, is built on them. At the foot of a cliff, where the full force of the wind is likely to be felt, a strong structure of masonry is usually built, while on the top of a cliff an ordinary brick or stone tower is sufficient. This latter form of lighthouse is usually connected with the keepers' rooms, which, in this case, form a separate structure.

There are many parts of our coasts where some warning or guiding light is needed, but where a lighthouse would be, if not an impossibility, a very risky and costly undertaking. In such places a floating instead of a fixed light

Eddystone Lighthouse

is used, and is made to take the form of a real vessel
which may act as a guide and a warning to mariners, so
that ships may be kept from dangerous shoals and rocks.
Such a vessel must be of fine sea qualities, for it has
to remain in the neighbourhood of the dangerous spot
throughout the year, in fair weather or foul.

A lightship is very carefully designed and its massive
iron sides are bolstered and braced together with bands
of steel or iron, and the high bows make the vessel almost
invulnerable in the strongest gales. The lightship is

The Abertay Lightship

provided with exceedingly reliable moorings. Massive
cables and anchors are attached for the purpose of meeting
the enormous strain to which they are subjected at times
of violent storms. The chain cables are tested to a
breaking strain of sixty tons, and they are frequently
examined.

Twelve men usually form the crew of a light-vessel
and seven men at least must be on board at any given
time. The violent pitching and tossing of a lightship in
bad weather is appalling, and sometimes for days together

there is no running for shelter, but only a dreary spell of waiting for fine weather.

At night time the great circular lantern is hoisted up the mast to a height of perhaps fifty feet above the deck. The lights of all lightships differ from one another so as to prevent any mistakes. The colours of some are red, of others green, but the greatest number are bright, either fixed or occulting, in some form or other. The illuminant is oil, and in most cases paraffin is used, being stored on board in enormous tanks.

In addition to the light from lighthouses and lightships, provision has to be made for giving warning in foggy weather, when a light is of little use. This takes the form of a syren, which is sounded by compressed air or steam, and the duration or variation of the note or blast signifies the position and name of the danger. When a vessel is seen in danger from the lightship, a warning gun is fired. Signals in the form of guns and rockets are also fired to give notice to the shore, and so call the lifeboat, when a vessel is wrecked.

Certain lightships are in telephonic communication with the land, and telegrams can also be despatched from them. Submarine signalling apparatus is also being installed in them, so that the efficiency of the modern lightship is constantly improving.

14. THE PORT OF LONDON

London has always stood first among British seaports, and for the last two hundred years it has been the greatest port in the world. The reasons for this premier position are chiefly because of its geographical advantages. It is situated on both sides of the Thames, a broad navigable river, running far into the land, and having a tide which twice a day carries vessels into its docks. It stands opposite the great markets and chief ports of the Continent, and this nearness to the continental ports is probably the main cause of its large *entrepôt* trade. The great value of its imports and the extent of its *entrepôt* trade are the special features of its foreign trade. With regard to the value of its exports it is surpassed by Liverpool, owing to the more favourable situation of the latter port with reference to the great manufacturing districts of England.

The important position which the Port of London occupies in relation to the foreign trade of the country is indicated by the following figures, showing the value of the imports and exports (excluding coastwise goods) of the United Kingdom as a whole and of London in particular:

United Kingdom   ..   £1,343,601,761
London £383,629,052

For some years before 1908 there had been a steadily growing feeling that London was not keeping pace with its competitors, or making proper provision for the future development of the shipping trade. Various proposals were made to improve the Port of London, and as the result of years of discussion a Bill was passed for the "improvement and better administration of the Port of London." The Port of London Authority came into existence on March 31st, 1909, when it took over the docks, and has done its best to administer, preserve, and improve the Port of London.

The authority of the Port of London extends from Teddington to a line drawn from Havengore Creek in Essex to Warden Point in Sheppey, a distance of nearly 70 miles. The upper portion of this section, between Teddington and London, is mainly used for boats, river sports, and generally as a pleasure resort, the centres of attraction being Twickenham, Richmond, Kew, Hammersmith, and Putney, whilst in and below London the riverside is lined with wharves engaged in almost every branch of trade connected with the commerce of the Port.

With the exception of the Surrey Commercial Docks on the south side of the Thames, all the other docks are situated on the north side of the river and are as follows: St Katharine's, London, West India, Millwall, East India, Royal Victoria, Royal Albert, and Tilbury Docks. From St Katharine's Docks near the Tower of London it is twenty-two miles to the newest and most modern deep-water docks at Tilbury which were opened in 1886.

The water area of these docks is 746 acres; the warehouses and other buildings occupy another 2178 acres; and the quay accommodation of the Port of London extends to forty-five miles. The estate of the Port of London may be said to include various systems of wet docks and dry docks, together with warehouses, granaries, elevators, and extensive machinery and plant.

Before we pass in review the various docks it may be well to explain that about sixty per cent. of the shipping entering the Port of London discharges in the docks, while the rest discharges at wharves or at moorings in the river. From vessels entering the docks a large part of the cargo is taken away by lighters, another portion is discharged on to the dock quays for conveyance to its destination by land, and yet another part is stored in the Port Authority's warehouses. This warehousing business is of great importance, for it comprises every class of goods entering the port. The principal imports are grain, timber, wool, frozen and chilled meat, sugar, tea, wines and spirits, and tobacco. The whole of the tobacco warehoused in London is stored with the Port Authority.

The St Katharine Docks adjoin the Tower Bridge and occupy the site of the St Katharine Hospital, an ancient charitable foundation. Steamers of moderate size enter these Docks and the warehouses are filled with some of the most valuable articles of produce coming to the Port of London. Here are stored immense quantities of tea, indigo, scent, wool, and shells, especially mother-of-pearl and tortoise shell.

The London Docks are quite close to the last-named docks and are the home of a most important section of the continental and coasting trades, and they warehouse produce from all parts of the world. The storage capacity of the warehouses and vaults in these docks is of vast extent. There is storing accommodation on the various floors for 220,000 tons of goods, and the vaults have room for 105,000 pipes of wine. Produce from every part of the world may be seen in the warehouses, and among the valuable goods are wool, wine, brandy, sugar, rubber, gutta percha, ivory, spices, drugs, coffee, cocoa, isinglass, and quicksilver. The warehouse storing the ivory is of great interest. The bulk of this article comes from Africa, India, Ceylon, and Siberia, but that from Africa is superior in whiteness and density to every other description. The warehouses storing the wool have yearly about 1,500,000 bales of a value estimated at £20,000,000; while the accommodation for the storage of rubber is the largest in the kingdom.

The Surrey Commercial Docks have a vast trade in timber and grain and are the chief centre for the importation of Canadian produce. The grain received at these docks comes mainly from the Baltic, the Black Sea, Canada, the United States, and the River Plate, and is discharged from vessels into craft or alongside granaries either with elevators, or by corn porters. The principal granaries have machinery for transferring the grain direct from the ship's hold or craft on to band machinery, from which it is distributed on any floor and to any part of the warehouse desired.

The Surrey Commercial Docks are the largest emporium for wood goods in the world. Besides the enormous sheds for storing the timber, there are large ponds for loads of floated timber. The greater proportion of the timber in these docks is soft wood coming from the Baltic, the White Sea, the United States, and Canada. The quantities of hard wood chiefly used in cabinet-making and for other purposes come from South America in liners.

The Surrey Commercial Docks have the best accommodation for Canadian produce, which is brought over in refrigerated steamers and transferred from the ships in a few minutes to the cold storage in these docks. Besides holding immense quantities of Canadian cheese and bacon, the cold storage provision is equally suitable for frozen meat and other articles from other countries.

The West India Docks are in the northern part of the Isle of Dogs, Poplar. There is extensive warehousing accommodation at these docks for rum, sugar, hops, grain, seed, and various descriptions of wood, especially mahogany, teak, ebony, satin-wood, and lignum vitae. In one warehouse there is storage capacity for 90,000 carcases of sheep which come from Australia, New Zealand, and Argentina. Some of the largest ships carrying East India and colonial cargoes discharge in this dock, and most of the large sailing ships coming to the Port of London load here.

The East India Docks are about half a mile east of the last docks and are principally used by sailing ships and by the fine steamers of the Union-Castle Line trading to the Cape and East Africa.

The Millwall Dock is in the Isle of Dogs, south of the West India Docks. Grain handling is a special feature of this dock, and it is estimated that two-fifths of the grain coming into the Port of London is delivered into its warehouses. The Central Granary is a huge building of thirteen floors built for dealing expeditiously with the discharge and housing of grain cargoes principally from the Baltic and the Black Sea.

The Royal Victoria and Albert Docks are the largest system of docks under the control of the Port Authority. The warehousing business is principally in grain, tobacco, and frozen meat. The tobacco warehouses have from 15,000 to 20,000 tons of tobacco in bond at one time, and are the only depôt of the kind in London. The stores for the frozen meat business are very extensive, and perhaps the largest in the world. There are at one time in cold

East India Docks

storage as many as 1,000,000 carcases of frozen mutton besides thousands of quarters of chilled or frozen beef. In these docks may be seen many of the finest steamers coming from China, the East Indies, Australia, New Zealand, and South America. The largest line is that of the Peninsular and Oriental Steam Navigation Company, whose ships have the carrying of the mails to the East Indies and Australia.

The Tilbury Docks are the farthest from London, but are in direct railway communication with all parts of Great Britain. The dry docks are the largest in the Port of London and there is a steam floating crane capable of lifting fifty tons. In no other part of the London docks is to be seen such a goodly array of powerful modern liners, many having a register of 12,000 to 14,000 tons. Special facilities are afforded at these docks for part cargo vessels, such ships not requiring the full benefit of enclosed docks but simply the unloading at riverside berths on the Thames.

15. GREAT SEAPORTS AND THEIR INDUSTRIES

It will be gathered from the preceding chapter that London is a great market port and serves a population of about 7,000,000 in and around the city. To some extent Harwich, Queenborough, the Channel ports, and even Southampton are involved in the vast trade of London. Of its whole trade nearly seventy per cent. is import; and the largest portion of the goods brought into the docks from overseas is at once carried away to wharf or warehouse. London, Liverpool, and Hull are essentially ports of lock-closed docks, where the water remains impounded at high-tide level. Glasgow, on the other hand, is a good example of an open or tidal port. We will now proceed to notice briefly these and some other British ports which have risen into importance owing to their situation, range of tide, barge facilities, and connexion with the great railway lines.

Liverpool landing stage

Liverpool ranks as a seaport next to London in importance. It is the principal port, not only for the industrial region of south Lancashire, but for many parts of the Midlands as well. Liverpool's trade has always been with America, and in early times it had a brisk trade with that country in negro slaves. It was, however, the rise of the English cotton trade and the rapid growth of so many large towns in Lancashire that made Liverpool surpass all its rivals except London. Prompt despatch is the order for the port, and there is a constant increase in the dock and railway facilities. Its twenty-five miles of quays have ships moored to them from every corner of the globe. Most of the principal steamship lines, including the Cunard and the White Star, have sailings from Liverpool, and although some of its ocean passenger traffic has been diverted to other ports, the growth of its goods trade is still more profitable.

Manchester has become a great seaport since the opening of the Ship Canal and is consequently a rival of Liverpool. The tonnage is not so great as in many other ports, but the value of the imported and exported goods is high. While the Canal has not been a great financial success, there is no doubt that the trade interests have been benefited by its construction.

The ports of Hull, Grimsby, and Goole constitute the Humber ports. Hull is the third port of the United Kingdom, thus ranking after Liverpool. It has a large trade with all the countries of Western Europe, and like Grimsby it is a convenient port for the North Sea fishing banks. Hull imports large quantities of corn and grain, oil seeds, butter, and timber, and exports machinery, cotton goods, and coal.

Southampton is the chief seaport on the south coast of England and the fifth port of the United Kingdom. It has grown rapidly and has many advantages for trade, for it is favourably situated on the busiest commercial sea route in the world—that between the English Channel and New York. It has a great advantage in being near London, in proximity to the Continent, and in having four tides daily instead of two. This port has obtained most of the trade with South America and South Africa and is a great calling-port for ocean liners.

Glasgow is the sixth port in the United Kingdom and the first port in Scotland. Great sums of money have been spent in making the Clyde navigable at Glasgow. A century and a half ago people could wade across the river where now great ocean liners sail in safety. Above Glasgow the Clyde is of no importance as a commercial highway, but from Glasgow to the mouth of the river the banks of the Clyde are lined with shipbuilding towns that make this the greatest shipbuilding district in the world. Glasgow Harbour is four and a half miles in length and has 304 acres of water space. There are several docks and the quays are ten miles long.

Bristol is one of the oldest seaports of Britain and was formerly a port of great renown, having a large trade with the West Indies, America, and Ireland. The trade of Bristol is still carried on chiefly with the two latter countries. Tobacco and raw cocoa are imported in large quantities, and from the West Indies come sugar, bananas, and pineapples. The huge ships of to-day are able to come up the Avon to Bristol, and so great docks and an "out-port" have been constructed to meet the increasing trade.

Harwich is one of the most important ports on the east coast. It is the best natural harbour between the Thames and the Humber, and it can be reached in an hour and a half from London. The harbour is formed by the estuaries of the Orwell and the Stour, and the most important development of the place is that of Parkeston Quay, owned by the Great Eastern Railway, whose steamers run daily from here to the Hook of Holland, Antwerp, Rotterdam, Esbjerg, and Gothenburg. There is not much export trade at Harwich, but there is a large import trade of perishable articles such as butter, eggs, meat, fish, and poultry, and of high grade fancy goods from the Continent.

Leith is the second port of Scotland and is the outlet for seaborne traffic from Edinburgh. Most of its trade is with the Continent, especially the ports of the Baltic

Newcastle-on-Tyne

and the North Sea. Corn, grain, and provisions of all kinds are imported, and machinery, iron, hardware, fish, and spirits are among the chief exports.

The Tyne ports comprise Newcastle, and North and South Shields. The port authority is the Tyne Improvement Commission who have spent about £7,000,000 in deepening, straightening, and widening the river, and in the construction of piers, docks, and wharves. The Tyne is now the most important shipbuilding river in England

and the coal trade of Newcastle is proverbial. The imports are mainly food stuffs, spirits, wines and timber.

The other north-east coast ports are those on the river Tees, which include Middlesbrough, the Hartlepools, and Stockton, and Sunderland on the Wear. Like Newcastle they are all engaged in shipbuilding, coal is exported, machinery is made, and locomotives are built. Middlesbrough is the chief town in the Cleveland district, so rich in iron ore. Less than a century ago there was only one house where now there is a great town with more than 100,000 people.

The most important Welsh ports are Cardiff, Swansea, and Newport, all engaged in the coal trade. Cardiff, including Barry Dock and Penarth, is the first port in Wales, and the largest exporter of coal in the kingdom, the second timber port, and comes next to London and Liverpool in the import of chilled and frozen meat. The coal is the finest in the world and is largely used by the British Admiralty. Swansea is the most important metal-smelting town in Britain, and the largest centre in the world of the tin-plate industry.

Newhaven, Folkestone, and Dover are the principal ports for passage to France, having daily services of fast steamers to Dieppe, Boulogne, and Calais. Their import and export trade is large, the import of silk being very considerable, especially at Newhaven and Folkestone.

Dublin, Belfast, and Cork are the three largest sea-ports in Ireland. Dublin is favourably situated for trade, for from it the railways radiate to all parts of Ireland, and its bay is the best harbour in the neighbourhood. Here, too, the Irish Sea narrows to sixty miles, and from Kingstown, six miles down the bay, there is fast mail and passenger traffic to Holyhead.

Belfast is a great shipbuilding centre, and claims to have the largest shipyard, the largest linen-mill, the largest mineral-water factory, the largest rope-work, and the largest tobacco factory in the world. Most of the trade of Belfast is with Glasgow and Liverpool.

Cork owes much of its prosperity to its harbour. Queenstown, its out-port, is the most important calling place for Atlantic liners in Ireland. The mails from America are landed at Queenstown and sent by rail to Kingstown. Thence they are despatched by cross-channel steamers to Holyhead, and then carried by express trains to London and other large centres. Cork has the reputation of exporting more dairy produce than any other town in the British Isles.

16. THE CINQUE PORTS

We shall understand the story of the Cinque Ports much better if we first give a little attention to the nature of the sea-coast on the south-eastern shores of England. It has been said that these shores have been more altered in form, more cut away in projecting parts, and more filled up in the recesses, than any other parts of the British coasts. This change has been brought about by the constant action of the winds and tides upon the materials within their reach; and one may observe masses of shingle and sand lining the Kent and Sussex coasts, which have been deposited by these agencies upon strata of a kind different from themselves. Hence it is safe to conclude that these masses of shingle and sand, swept along by the force of the south-west gales, acting during the flood tide, are the agents that have closed the harbours, choked the ports, and changed the form of the coast. The wind and tide produce these effects in so uniform a manner, that their action has been described as the result of a natural law—the law of Eastward Drift.

The operation of this law may be traced first at Hastings, then at Winchelsea, Rye, Romney, Lydd, Hythe, Folkestone, Dover, Walmer, and Deal, and lastly at Sandwich and the ports in its neighbourhood. All have been separated from the sea and although Dover appears an exception, it has only held its own by the aid of vast sums from the national exchequer.

We are now in a better position to consider the Cinque Ports which once formed a flourishing and powerful confederation. Originally there were five ports, hence the French name, but a few years after the Conquest two others were added to the confederation, which is officially known as "the Five Cinque Ports and Two Ancient Towns." The Cinque Ports are Hastings, Sandwich, Romney, Hythe, and Dover, and the two Ancient Towns are Winchelsea and Rye. Besides these seven ports, almost every town from Pevensey in Sussex to Faversham in Kent was attached to the Cinque Ports as "limbs" and some of these "limbs," such as Tenterden, were far from the sea.

Now what was the origin of this confederation of the Cinque Ports? Some writers are inclined to the view that they were five of the fortresses which the Count of the Saxon shore had under his control to guard the landing places of the Narrow Seas from Yarmouth to Portsmouth. Other authorities trace the origin of the Cinque Ports to a later period when the Teutonic invaders had settled in our land. Whatever their origin, these

Cinque Ports have played an important part in our history and from many points of view are worthy of further consideration.

The Cinque Ports had various duties to fulfil, and in return their privileges were considerable. Their most important duty was to guard the south-eastern coast and to supply shipping in defence of the realm. In the reign of Henry III they had to provide fifty-seven ships, each

The Cinque Ports

with twenty-one men and one boy, to serve the King at their own cost for fifteen days, and thereafter as long as required for a special payment. Dover had to find twenty-one ships, Winchelsea ten, Hastings six, and Sandwich, Hythe, Romney, and Rye five each. These ships engaged in a naval battle when the fleet of Lewis of France was utterly defeated and England was saved from French invasion. It has been well said "that the courage of those sailors who manned the rude barks of the Cinque Ports first made the flag of England terrible on the seas." The sailors were thanked by the King's Council for their frequent and excellent services both to the King and his ancestors; and henceforward we frequently find references to the Ports as supplying ships for conveying troops and royal personages to the Continent and performing other maritime business.

The privileges granted to the Cinque Ports were very valuable. The ports were all self-governed or free towns, and their freemen were allowed to trade free of toll either at home or elsewhere in the kingdom. The men were exempt from all military duty, and whatever offences were committed by the people of the Ports could only be tried by the Lord Warden who was the chief man of the confederation, or by the King. The Cinque Ports had their own courts, the chief being the Court of Brotherhood, which was held twice a year, once at Shepway Cross near Hythe, and then at Romney. This court was attended by seven persons from each port, and the members met to uphold their privileges and regulate their own affairs.

Arms of the
Cinque Ports

Down to recent times the Cinque Ports had a large representation in the House of Commons; and the old custom of carrying canopies of gold or silk over the King and Queen at their coronation is still retained. The Barons of the Cinque Ports claim to sit at the right hand of the King at the banquet, and afterwards divide the canopies among themselves as their fee.

The arms of the Cinque Ports are three half lions and three half ships. In the early days the sailors wore a special dress, which consisted of a white cotton coat with a red cross, and the arms of the ports underneath.

The Cinque Ports were at their prime in the early Plantagenet period of our history, and for more than a century the Royal Navy of the Cinque Ports with its distinctive flags and banners took part in many engagements. In the reign of Edward III, however, the disasters of the Ports began, and from them they never recovered. The decay of the fine harbours of the Cinque Ports was the main cause of this decline, and when Henry VII began to form his Royal Navy the assistance of the Cinque Ports was no longer needed. Although the work of the Cinque Ports has been long finished they still retain some of their old privileges and customs. Their chief officer, the Lord Warden, who is often one of our great statesmen, presides over their courts. The Duke of Wellington was very proud of his Lord Wardenship and he spent his last days at Walmer Castle. Since his death the office has been filled by such famous men as Lord Palmerston and Lord Salisbury, and, for a time, our King George V, when Prince of Wales, gave royal dignity to this ancient office.

It will now be gathered that the Cinque Ports declined through no shortcoming of their own but owing to physical and political changes. They have played a notable part in the history of our country for they helped to secure the sovereignty of Great Britain in the Narrow Seas and their fleet has been called the parent of our Royal Navy.

We will now briefly consider each of the Cinque Ports and Ancient Towns. It is well known that the present is not the original town of Hastings, which occupied a site that has been long covered by the sea. In ancient days Hastings had a good harbour and a pier; and a little to the west of the ruins of the castle there was formerly a great priory of the Black Canons founded in the reign of Richard I.

Sandwich was once the resort of vessels of all sizes from many quarters. The action of the tide silted up the harbour, and no attempt was made to save it. Now the town is nearly two miles from the sea, or four miles by the winding course of the Stour. Sandwich presents a perfect picture of a port which the sea has forsaken, rising above flat marshes reclaimed from the sea. The town, however, retains some marks of its former importance, and the Fisher Gate, St Clement's Church, and St Peter's Church are of historic interest. The most picturesque object in Sandwich is, however, the old bridge, formerly a drawbridge, having its Barbican and flanking towers, with the Fisher Gate grooved for portcullis near at hand.

Of Romney there is little to be said, for it has no harbour, standing as it does a mile from the sea. It has little trade, but is celebrated for its great sheep fair in August.

Hythe is another port from which the sea has retreated. It is now a clean, well-ordered town, with a bank of shingle one mile in width before the sea is reached. Hythe is well known for the Government School of Musketry and its rifle-butts.

In every way Dover is the most important of the Cinque Ports. Although it has not a good natural harbour, large sums have been spent to improve its accommodation. Its trade chiefly depends on the continental service, and it is the terminus of the South-Eastern and Chatham Railway. Dover has shipbuilding, rope and sail-making, fisheries, and some coast traffic. The harbour consists of two docks and a tidal basin, and the Admiralty Pier, 600 feet long, forms one side of a natural harbour of refuge. This harbour with a water area of over 600 acres will take the largest vessels and has cost about £4,000,000. The continental mail-boats depart from the Admiralty Pier, on to which the boat trains run.

Winchelsea has suffered more than the other ports in Sussex. The old town lies beneath the sea, and the new Winchelsea, built by Edward I, has been left a mile inland. The lost city was at the height of its glory in the reign of John, when its bay was the place of rendezvous for the fleets of England, and its commerce was great and flourishing. Old Winchelsea was, however, doomed to be destroyed, and during the first half of the thirteenth century mighty storms occurred, so that Winchelsea was inundated in 1236. Year by year the town suffered more inroads by the sea, until in 1287 the catastrophe came which annihilated the town, whose streets, churches, and dwellings were covered by the sea. Not far from the site of Old Winchelsea was a little promontory called Iham, secure from the fury of the sea, and there land was bought and the New Winchelsea was built. On all sides save one it was surrounded by the sea; there was good harbourage; and defence was easy. The new town was well planned, the streets were wide and open, there were strong walls on the land side, and three gates, which still remain, gave entrance to the town.

New Winchelsea soon rose to a high position as a Cinque Port, and its men and ships were the backbone of the fleets sent against the enemy. It had its disasters, for it was sacked and burned by the French, and its walls battered down. But a worse fate was in store for it, for the sea which overwhelmed Old Winchelsea receded from New Winchelsea, leaving it to the fate of a stranded port. By the middle of the fifteenth century its sea trade had come to an end, and its merchants had left the town for other ports. The subsequent history of Winchelsea is not of importance, but it derived some notoriety in the eighteenth century from smuggling, which brought its residents some prosperity. This illegal trade flourished and was winked at by magistrates, squires, and parsons. Thackeray in *Denis Duval* has immortalised two respectable country gentlemen who were connected with this profitable trade and who were executed at Tyburn.

Rye

Of all the seaports which the tide has left high and dry the most picturesque is Rye. It is at the mouth of the Rother and was once a famous seaport, but owing to the changes in the coast-line the town is now two miles from the sea. In the fifteenth century Rye combined with Yarmouth in the fishing trade in the North Sea, and its fleet engaged in the wine, timber, and billet trade.

Its former importance is shown by the concluding words of its charter: "God save Englande and the towne of Rye"; and by the fact that when Queen Elizabeth visited it in 1573 she named it "Rye Royal." One of the most ancient buildings in Rye is the Ypres Tower, long used as a prison and a watch tower. It stands on the cliff, and was built in Stephen's reign by William of Ypres, Earl of Kent. It is square in plan, with round towers at the angles, and was for some time the only defence of the town.

INDEX

Milton Keynes UK
Ingram Content Group UK Ltd.
UKHW032320161024
449665UK00001B/18